Is an Elect
Right for You?

Quick Reference Guide to Buying an Electric Vehicle

Is an Electric Car Right for You?

Quick Reference Guide to Buying an Electric Vehicle

CHRIS JOHNSTON

ED SOBEY

Warrendale, Pennsylvania, USA

400 Commonwealth Drive
Warrendale, PA 15096-0001 USA
E-mail: CustomerService@sae.org
Phone: 877-606-7323 (inside USA and Canada)
724-776-4970 (outside USA)
FAX: 724-776-0790

Library of Congress Catalog Number 2022942597
http://dx.doi.org/10.4271/9781468604863

ISBN-Print 978-1-4686-0486-3
ISBN-PDF 978-1-4686-0487-0
ISBN-epub 978-1-4686-0488-7

To purchase bulk quantities, please contact: SAE Customer Service

E-mail: CustomerService@sae.org
Phone: 877-606-7323 (inside USA and Canada)
724-776-4970 (outside USA)
Fax: 724-776-0790

Visit the SAE International Bookstore at books.sae.org

Publisher
Sherry Dickinson Nigam

Development Editor
Amanda Zeidan

Director of Content Management
Kelli Zilko

Production and Manufacturing Associate
Brandon Joy

Contents

Foreword

The Economic Advantages of Going All-Electric by Erika Myers

I often hear other women express their frustration and uncertainty when looking for a more fuel-efficient, cleaner, and environmentally friendly car. They ask, "Which is better in the long run—a traditional hybrid (HEV), plug-in hybrid (PHEV), or battery electric vehicle (BEV)?"

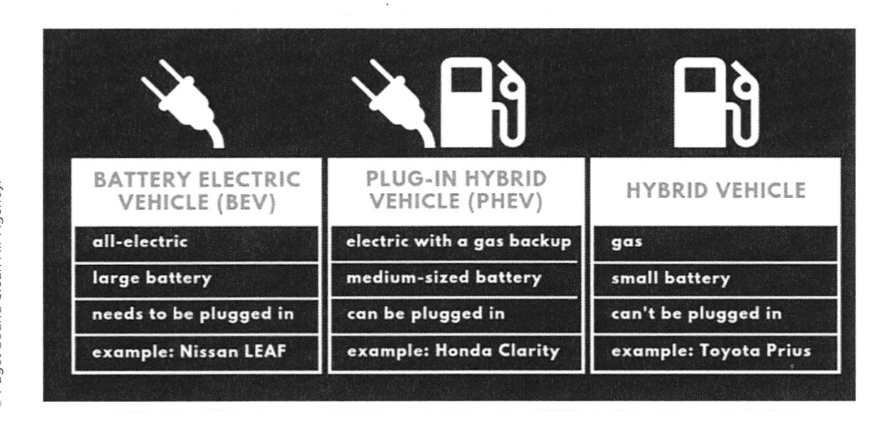

© Puget Sound Clean Air Agency.

When it comes down to it, there are cost savings for transitioning to ANY of these options compared to a traditional gas-only fueled engine. However, if you're going to make the switch to a vehicle with an electric drivetrain, the fully electric vehicle (EV) has economic advantages compared to traditional and plug-in hybrid models. Here's why.

EVs Cost Less to Maintain

First and foremost, it's as simple as "less is more." Why have two propulsion systems in your vehicle when you can have just one? Consumer Reports dug into the maintenance of EVs, PHEVs, and internal combustion engine (ICE) vehicles and found that EV and PHEV drivers spend half as much as ICE owners to maintain and repair their vehicles, saving an average of $4,600 in repair and maintenance costs over its lifetime. [1]

All-Electric Repairs Are Easier than Plug-In and Traditional Hybrids

PHEVs and HEVs still have an ICE, thus requiring essentially the same maintenance as ICE vehicles, even though the engine is used less.

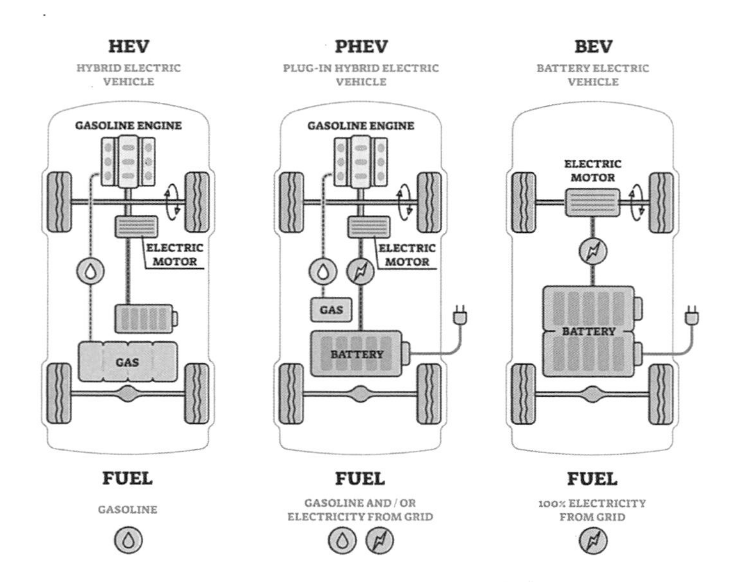

VectorMine/Shutterstock.com.

[1] https://www.consumerreports.org/car-repair-maintenance/pay-less-for-vehicle-maintenance-with-an-ev/

The major repair of an EV is replacing its battery pack. A battery that is being used loses capacity over time, thus reducing your maximum range. At some point you'll want to replace the battery pack, but you can continue to operate the vehicle risk-free until then. With PHEV, HEV, and ICE vehicles, engine and transmission problems can happen suddenly and are more likely to cause catastrophic damage or leave the combustion engine inoperable until repaired.

EV Prices and Models Are More Competitive than Ever

The EV market is expanding rapidly, meaning that there are even more opportunities to purchase different-sized vehicles and even used EVs. Newer EV models now offer longer battery ranges than their predecessors and are rapidly lowering in price as battery manufacturing costs have declined nearly 80% since 2013.[2]

For example, here's a comparison of Hyundai's 2021 models (Table 1).[3] With tax credits factored in, the electric and plug-in models are comparable—even less expensive—than the traditional hybrid models.

TABLE 1 Comparison of Hyundai's 2021 models.

Model	Engine type	Price (USD)	All-electric range
2021 Sonata Hybrid	Traditional hybrid	$27,750	N/A
2021 IONIQ Plug-In Hybrid	Plug-in hybrid	$22,157*	29 miles
2021 IONIQ Electric	Electric vehicle	$25,745*	170 miles
2021 Sonata Electric	Electric vehicle	$29,890*	258 miles

* Price calculated after federal tax credit.

Courtesy of Erica Myers.

If you have specific models in mind, the Department of Energy's Vehicle Cost Calculator (https://afdc.energy.gov/calc/) is a great way

[2] https://about.bnef.com/blog/battery-pack-prices-cited-below-100-kwh-for-the-first-time-in-2020-while-market-average-sits-at-137-kwh/#:~:text=By%202023%2C%20average%20prices%20will,%24100%2FkWh%20have%20been%20reported

[3] https://www.hyundaiusa.com/us/en/hybrid-electric

to compare costs using your daily driving distance. You can also see a comprehensive list of all 2021 EVs and PHEVs at EVAdoption (https://evadoption.com/ev-models/bev-models-currently-available-in-the-us/) and at (https://evadoption.com/ev-models/available-phevs/#), respectively.

EVs Fuel Costs Are Consistent and Less Expensive

You may have noticed over the past few months that gas prices are going up. Gasoline prices are subject to supply and demand, which can be affected by everything from weather events to geopolitical events. While HEVs and PHEVs are still tethered to the gas pump, EVs are divorced from the gas price fluctuation headaches because electricity prices are far more predictable and stable over time. Last year the National Renewable Energy Laboratory found that, depending on which state you live in, you can save up to $14,500 in fuel savings over the lifetime of an EV when compared to ICE vehicles.[4]

That being said, there is an up-front cost of purchasing and installing a Level 2 charger in your home if you want a quicker charge. These estimates are between $1,000 and $2,900 depending on your setup and features. While these up-front equipment and installation costs are significant, there are federal, state, and local discounts, and utility and charging companies who can often provide discounts, financing, tax credits, and rebates to help.

Overall, buying an EV is not just a great decision. It's also one that will benefit your pocketbook for years to come.

Erika Myers
Founder, EV Love
https://electricvehiclelove.com

[4] https://www.nrel.gov/news/press/2020/research-determines-financial-benefit-from-driving-electric-vehicles.html

EV Love is a site dedicated to electrifying mobility one woman at a time.

Courtesy of Erica Myers.

Jeerasak banditram/Shutterstock

Introduction

We're stoked about electric vehicles (EVs). So is the car industry. The industry is transitioning from producing gasoline and diesel-powered cars to cars powered by electricity. In a few years some of the manufacturers may not even be making gasoline-powered cars. If your next car isn't electric, the car after that undoubtedly will be!

Start with these basic questions to assess your overall needs. The rest of this quick reference is dedicated to answering your questions about what is an electric car and how do they work, explaining what you need to know before you buy, and, finally, a section dedicated to understanding the truth behind common misconceptions.

FIGURE I.1 EV sales accounted for 73% of all plug-in EV sales in 2021.[1]

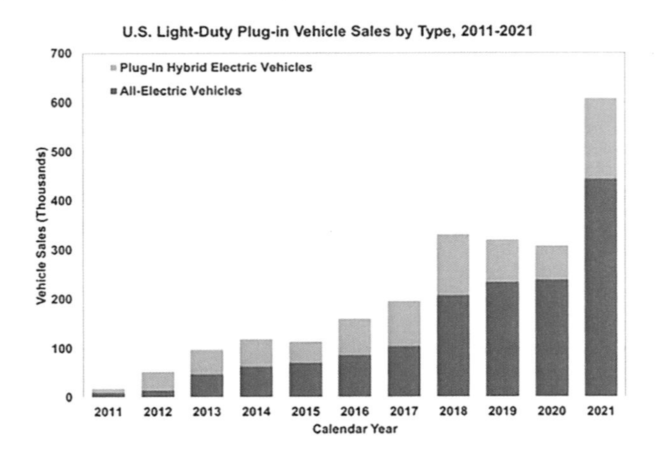

Courtesy of Energy.gov.

[1] https://www.energy.gov/energysaver/articles/new-plug-electric-vehicle-sales-united-states-nearly-doubled-2020-2021

Is an Electric Car the Right Choice for You?

We think EVs are the right choice for most drivers, but only you know what will work best for you. Here are a few considerations.

- How far do you drive every day?
- If your daily commute is more than 250 miles, only a few electric cars will meet your needs. For typical daily commutes, you can recharge the car battery overnight at home without having to stop and spend time at a charging station.
- Is the reliability of your car a major concern?
- If so, know that electric cars have vastly fewer moving parts in the drive system and require vastly fewer maintenance stops. You may never need a brake job for as long as you own an electric car and you won't need to change the antifreeze, have the transmission repaired, or replace the fuel or water pumps or change the spark plugs.
- Is driving safety an important issue for you?
- Electric cars are, in general, safer than internal combustion engine (ICE) cars. One reason is that electric cars don't have an engine mounted in front of the front seat passengers. In head-on collisions, electric cars crumple and absorb some of the shock. In a car with a front-mounted gas engine, that big engine gets shoved into the passenger compartment with little collision absorption.
- Is the environment a concern for you?
- There's a lot of inconsistent information floating about out there. Comparing apples to apples, electric cars pollute less and use less energy. Electric motors are more efficient than gasoline engines and they don't have an exhaust pipe. You benefit from that high efficiency by paying much less for the energy that moves your car down the road. Depending on the ever-fluctuating cost of gas, you can save $3 or $4 per gallon. If you currently drive around trying to save a nickel or dime per gallon, think about a savings of dollars per gallon.

Did you know in the first quarter of 2022, EV sales rose 76% over sales in the first quarter of 2021 [1].

FIGURE I.2 Author's car, a Tesla Model Y.

Courtesy of Ed Sobey.

1

The Revolution from Gasoline-Powered Vehicles to EVs Is Happening Now

It always seems impossible until it is done.

— *Nelson Mandela*

Why are we seeing an EV revolution today? EVs have finally overcome their three longstanding challenges of range, price, and styling. Today the automotive industry is at the cusp of a major inflection point to the extent that all major automotive manufacturers are "betting the farm" on EVs. For example, this is the first time in 55 years that Ford is using the Mustang badge (their most valuable) on a new car, and it's an EV (Figure 1.1). During Super Bowl 55, seven car manufacturers including General Motors, BMW, Kia, and Polestar spent a record $6.5 million for a thirty-second commercial about electric vehicles.

FIGURE 1.1 Ford Mustang Mach-E, all-electric SUV.

Mike Mareen/Shutterstock.com.

When it comes to EV advantages, you can argue that EVs are better in every way. With respect to performance, unlike a hybrid, an EV has instant acceleration. This not only makes them fast but very fun to drive as they lack the mechanical "lag" that's typical of gasoline-powered cars. EVs are very quiet, which also contributes to the enjoyable driving experience. It's easier to converse and the music sounds fantastic.

More EVs on the Road

Even more compelling, EVs accounted for 7.2% of global car sales in the first half of 2021, up from a meager 2.6% in 2019—a 177% increase in two years [2]. This uptake is unprecedented in the history of the automotive industry (Figure 1.2).

FIGURE 1.2 EVs and hybrids surpass 10% of United States (US) light-duty vehicle sales.

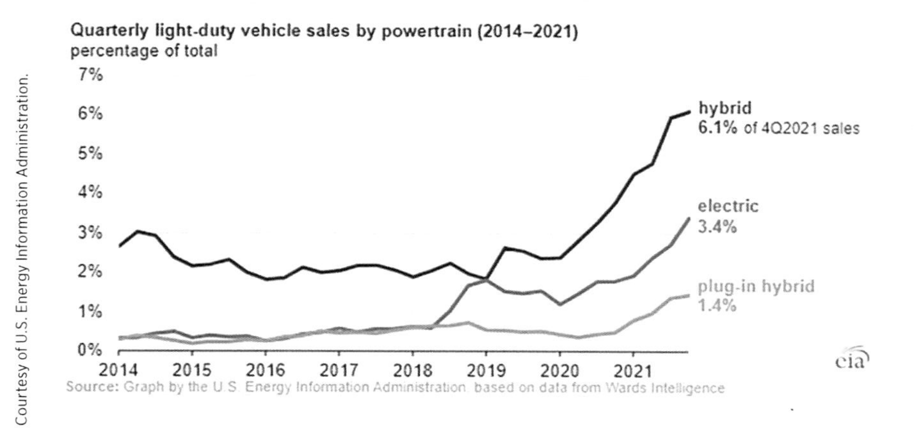

Innovative Design

Regarding design, EVs offer some of the most cutting-edge styling in the market, such as extra-large screens and more cargo storage in the "Frunk." With the absence of an ICE and a transmission, manufacturers are able to design vehicles differently, offering features that have not been seen before. For example, the Rivian R1T "gear tunnel" offers a slide-out platform to increase usable space and storage, and can even include products like the Camp Kitchen (Figure 1.3). The spacious interior of Hyundai IONIQ 5 allows for the front seats to completely recline. It even has leg rests and allows for the front passengers to enter and exit the vehicle from either side. Another example of an innovative design is the Lucid Air glass canopy. This optional solid glass roof spans the entire cabin almost without interruption, providing a spectacular view while blocking out heat and sunlight.

FIGURE 1.3 The Rivian R1T gear tunnel offers a slide-out platform for easy access storage and to increase usable space.

Roschetzky Photography/Shutterstock.com.

Did you know: When comparing the 22 mass-market EVs that are shipping in North America, the average range is 284 miles on one charge [3].

Unique Advantages

For the thrifty folks, an ICE vehicle will cost almost eight times more to operate and maintain than a comparable electric car. The very low cost of maintenance is accompanied by hassle-free ownership because of their high reliability (Table 1.1). EVs are the safest cars on the road because of several unparalleled design advantages, such as a lower center of gravity. When compared to gasoline or diesel-powered vehicles, EVs have a much lower carbon footprint, and they require much less carbon output to both manufacture and drive.

TABLE 1.1 Anticipated maintenance over a three-year ownership.

Battery electric vehicle	Internal combustion engine
Tires	Tires
	Oil changes
	Spark plug changes
	Transmission servicing
	Radiator fluids
	Oil filter
	Serpentine belt
	Brakes

Courtesy of Chris Johnston and Ed Sobey.

One of the advantages of EVs is that they have fewer components that require maintenance. Over a three-year period, owners of EVs might not have any maintenance required with the exception of replacing tires.

Land Speed Records and First Ladies: History of EVs

Even the history of EVs is fascinating. For example, the first six recorded land speed records of any vehicle were all held by EVs (Table 1.2). In 1899, an EV called the La Jamais Contente was the first vehicle of any kind to drive faster than 100 km/h (62 mph).

TABLE 1.2 First land speed records for any type of vehicle (Acheres, France).

Date	Vehicle	Engine	Driver	1 km (km/h)	1 mile (mph)
December 18, 1898	Jeantaud Duc Fulmen Battery Electric	Electric	Gaston de Chasseloup-Laubat	63.15	39.24
January 17, 1899	CGA Dogcart	Electric	Camille Jenatzy	66.66	41.42
January 17, 1899	Jeantaud Duc Fulmen Battery Electric	Electric	Gaston de Chasseloup-Laubat	70.31	43.69
January 27, 1899	CGA Dogcart	Electric	Camille Jenatzy	80.35	49.93
March 4, 1899	Jeantaud Duc Fulmen Battery Electric	Electric	Gaston de Chasseloup-Laubat	92.78	57.65
April 29, 1899	CITA No 25 La Jamais Contente Single action 4-cylinder	Electric	Camille Jenatzy	105.88	65.79

Courtesy of Chris Johnston and Ed Sobey.

Also interesting, starting in 1912, the Baker Electric Victoria was used by five First Ladies of the United States.

Courtesy of Newatlas.

1899—Belgian EV called the La Jamais Contente is the first vehicle of any type in history to go over 100 km/h (62 mph).

1912—Baker Electric Victoria, used by five first ladies of the United States.

2

What to Know Before You Buy

Whether you are considering purchasing your first EV or you are ready to buy, now is the time. There are more EVs on the market than ever before, and owning an EV has numerous advantages.

Performance

EV owners report that one of the most enjoyable features of driving their cars is the instant acceleration. Step on the accelerator and you are moving.

The faster acceleration is delivered by the torque that electric motors can deliver at low speeds. This is a significant advantage over gas-powered cars. Torque is the rotational force that causes a shaft to spin. Electric motors of EVs deliver 100% of their available torque instantaneously. They don't have to build up speed before they reach peak power. This enables them fast stoplight starts and superior passing ability.

FIGURE 2.1 Interior of a Subaru Solterra.

© Subaru of America.

An ICE power comes from a simple chemical reaction between fuel and oxygen. If you want more power, you must give it more fuel—hence the term "stepping on the gas." However, this isn't a quick process. The fuel needs to flow through pipes into an intake manifold and then finally into the combustion chambers. The hundreds of parts of the engine each have inertia that causes a lag when trying to get them to spin faster.

The biggest disadvantage that gas engines have with respect to torque output is their power band. The power band is the range of revolutions per minute (RPM) around peak power output. Electric motors usually have a flat power band, meaning that they deliver maximum torque the moment they start turning. ICEs are different. They must reach a high enough speed before they reach their power band.

Even if an ICE is capable of producing more power than an electric motor, it takes time to gain the RPM speed to hit its peak power. This is why EVs tend to accelerate much faster than ICE vehicles. The graphic below shows a rough comparison of the torque output from an electric motor versus a gas engine (Figure 2.2).

FIGURE 2.2 Graph showing a rough comparison of the torque output from an electric motor versus an ICE.

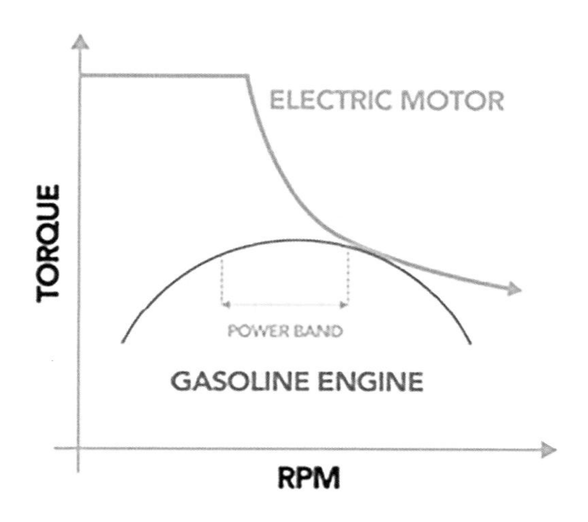

Courtesy of Chris Johnston and Ed Sobey

Low Cost of Ownership

Even though the price of EVs has steadily declined and will continue to do so, they are still priced at a modest premium over their ICE counterparts. However, when it comes to the cost of ownership, the data is very clear. EVs are less expensive to operate and maintain. We decided to compare costs over a five-year period.

VectorMine/Shutterstock.com.

Do you look at the price of gasoline every time you pull into a gas station? That's one of the larger costs of operating an ICE vehicle. If you drive as an average American does, you put about 10,500 miles on your odometer each year. Over five years, that is 52,500 miles. Taking a national average of 25.1 miles per gallon, that equates to 2,091 gallons of gasoline consumed over five years.

In April 2022, the national average price of gas was $4.13 for regular and $4.81 for premium [4]. For this comparison, we use the price of regular. That means that the typical gasoline-powered car owner will pay about $8,636 for gas over five years.

FIGURE 2.3 Retail price of gasoline in the United States from 1995 to 2020.

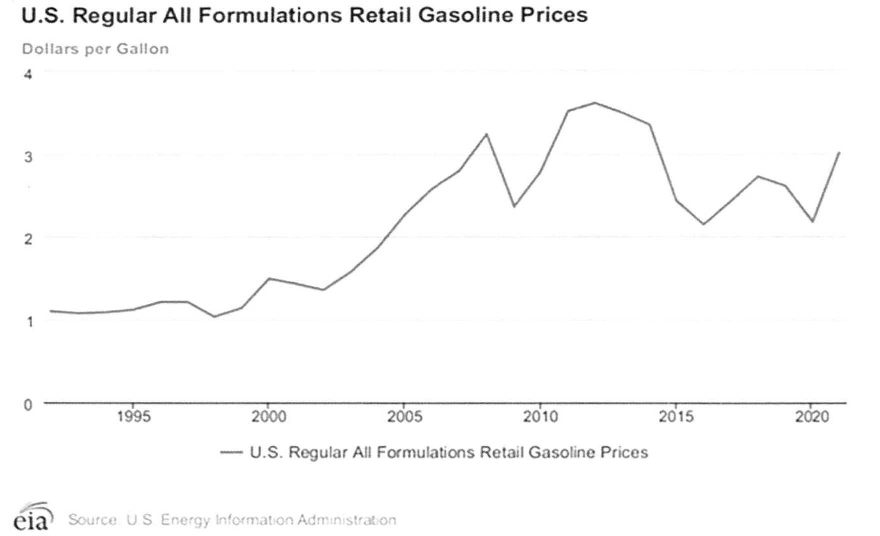

Courtesy of U.S. Energy Information Administration.

Determining the cost to charge an EV is trickier than determining the straightforward gasoline-consumption cost. The majority of the time, EV owners charge their vehicles at home or at work. For our simple comparison, we assumed that drivers charge their cars at home instead of using a public charging station. To determine at-home charging costs, we took the average cost of electricity for the home in the United States in 2021, which was

$0.1118 per kilowatt (kW) [5]. The average EV will use 10,900 kW to drive the American average of 52,500 miles [6]. That means that the typical EV owner will pay about $1,220 to charge their vehicle.

VectorMine/Shutterstock.com.

So, over a five-year period, the typical US EV driver saves $7,417 in the cost of fuel. The average EV driver also saves a ton of time and some aggravation by not having to go to gas stations.

FIGURE 2.4 Comparison of the power draw for typical home appliances.

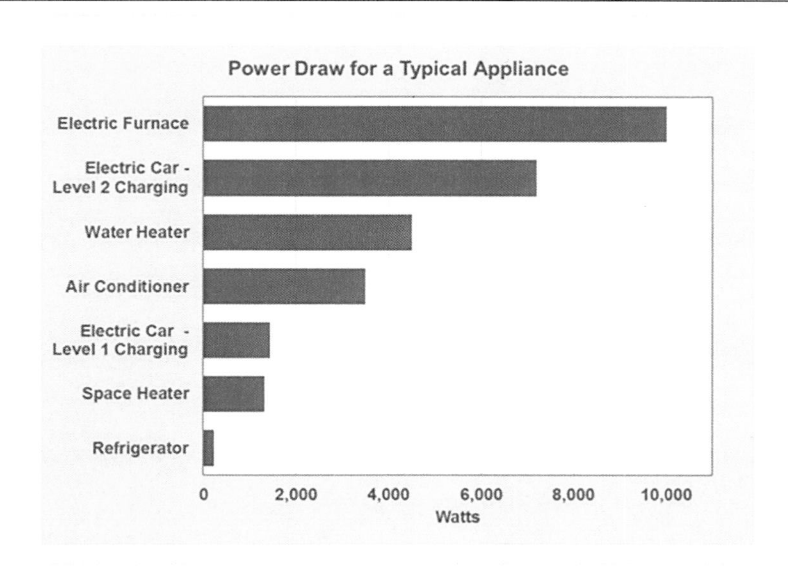

Courtesy of Energy.gov.

Did you know that the average home uses 30 kWh of power per day, so an EV battery could supply a home during a blackout for two to three days [7]?

Besides saving on fuel costs, EVs save money on brake maintenance. EVs save wear and tear on their brake pads and rotors by using the resistance of their motors to slow down and stop. This is called regenerative braking, and it drastically reduces the wear on the brakes.

EV owners can expect to get at least 150,000 miles from their brakes before they need servicing. On the other hand, gasoline-powered vehicles do not have regenerative braking. According to Kelley Blue Book (KBB), the average lifespan for brake pads on a gasoline-powered vehicle is about 40,000 miles. KBB states that the average cost to replace pads ranges between $150 and $300 per axle [8]. We chose a midpoint of $225 per axle. That means on average, ICE car drivers pay $450 every 40,000 miles to have all four brakes (two on each axle) serviced.

A third cost of operating a gasoline- or diesel-powered car is changing oil. The electric motors in EVs do not require oil. The American Automobile Association used to recommend changing oil every 3,000 miles. Newer synthetic lubricants extend this range to between 5,000 and 7,500 miles.

FIGURE 2.5 Commercial and private charging stations are coming online throughout the United States.

The Image Party/Shutterstock.com

For cost comparisons, we use an oil-change window of 7,500 miles. Thus, over our five-year operating-cost comparison between EVs and ICEs, the ICE owner would change oil seven times. KBB estimates that oil changes cost between $65 and $125 [9]. We use a midpoint cost of $95, which suggests a cost of $665 to change the oil seven times.

Here is a table summarizing the costs of fuel and maintenance over a five-year period (Table 2.1). Obviously, we are leaving out many potential repair bills that are impossible to predict. Of the maintenance items we know about, it is clear that EVs are much less expensive to operate. We estimate the savings at $4,815.

TABLE 2.1 Cost to operate and maintain over five years.

	Electric vehicle	Internal combustion engine
"Fuel"	$1,220 to charge	$8,636 for gas
Brakes	$0	$450
Oil	$0	$665
Total	$1,220	$9,751

Courtesy of Chris Johnston and Ed Sobey.

Safety

EVs are significantly safer than gas-powered vehicles for two reasons. First, because of their typical battery placement, EVs tend to have lower centers of gravity than gas-powered cars.

Having a low center of gravity makes an EV less likely to roll over. This is important because, according to the US Department of Transportation, rollovers have a higher fatality rate than other kinds of crashes [10]. With more weight below you in an EV, you are safer.

Second, a common cause of injury during a head-on collision is the ICE being pushed backward into the passenger compartment. The large block of metal has nowhere to go except into your lap.

An EV motor is much smaller and lighter than a gas or diesel engine. This has a few benefits. First, there is less heavy metal to be pushed back into the passenger compartment, causing injury. Second, EV motors are so small that they leave room for the manufacturers to put a trunk, or "frunk," in the front of the car.

Also, with the smaller electric motor, there is more empty space under the hood. When a crash occurs, that "crumple zone" will absorb much of the impact. The crumple zone acts as a shock absorber.

In 2021, the Tesla Model Y received a five-star rating in each category test by the National Highway Traffic Safety Administration including the lowest rollover risk ever recorded by the organization.

Reliability

Because of their engineering simplicity, EVs are far more reliable than gasoline-powered vehicles. It is difficult to get an exact parts count, but it is clear that EVs have fewer moving parts in their drivetrains than cars with ICEs. More parts, and especially more moving parts, mean more potential points of failure.

A fully electric car has far fewer moving parts, so they require less ongoing maintenance costs. In fact, there are about 20 moving parts in a typical EV drivetrain, compared to nearly 2,000 in an ICE vehicle [11].

While battery degradation varies by model and external conditions—such as climate and charging type—the majority of vehicles on the road today have not experienced a significant decline. In fact, overall degradation has been very modest, with an average capacity loss of just 2.3% per year. Under ideal climate and charging conditions, the loss is 1.6% [12].

VectorMine/Shutterstock.com.

Efficiency

Let's compare EV and ICE technologies in terms of efficiency. Efficiency is the amount of energy that you put into a vehicle as gasoline or electricity compared to how much energy you get out to move you forward.

The US Department of Energy found that EVs convert about 77% of received electrical energy from the grid to power at the wheels. Conventional gasoline vehicles only convert between 12% and 30% of the energy stored in gasoline to power the wheels [13].

This makes EVs two to three times more efficient than gas-powered vehicles. So, even if electric cars are recharged with electricity generated by fossil-fuel-based sources, which is true in some cases, their motors are more efficient at transferring that energy stored in the batteries into motion. That means much less waste and a better deal for the environment.

FIGURE 2.6 Polestar.

© Polestar.

3

Owning an EV

Owning an EV is a lot like owning a gas-powered vehicle, but some things are different. One is that you don't have to spend time at a gas station getting a fill-up. Of course, you can still stop to buy a bag of Doritos and be on your way without handling the gas pump.

Instead of filling up with gasoline, you need to charge your car. You can do that at home or at public charging stations. In this chapter, we cover a few items that you should consider when thinking about buying an EV.

Charging

EV charging comes in three "levels." Levels 1 and 2 can be performed from an owner's home or apartment. Level 3 requires a supply voltage that is typically only available in areas zoned as commercial or industrial. (Table 3.1).

Level 1 is the slowest way to charge an EV. It uses a standard 120-V wall outlet. Level 1 charging typically adds about 4 miles of range per hour. Depending on the model of EV, it can take between 8 and 24 hours to recharge the battery. Level 1 is convenient but slow. To charge faster, you need to use a higher voltage.

For peace of mind, keep a long extension cord in your trunk. We recommend a 100-foot, three-prong cord. That will allow you to recharge almost anywhere you go, even at campgrounds. As long as you can find a standard electrical outlet, you will be able to power your car.

TABLE 3.1 Comparison of the different levels of charging.

Charging level	Voltage required (V)	Locations	Charging time (miles of driving added per hour of charging)
1	120	Your home or business	4–5
2	240	Your home on a circuit wired for 240 V as is your clothes dryer and kitchen range.	18–24
3	480	Commercial charging stations, municipal and hotel parking lots	Fully charged

Courtesy of Chris Johnston and Ed Sobey.

Level 2 charging is typically what people install in their homes. It requires a 240 V electric circuit—the same as is used for large electric appliances like dryers or stoves. Level 2 charging is typically six times faster than Level 1. Level 2 adds about 25 miles of driving range per hour. Most EVs come with a portable Level 2 charger that includes a 240 V plug and a 120 V plug so that it can also be used as a Level 1 charger. If your clothes dryer is near your garage, you can probably use that outlet for charging your EV.

A slightly more convenient and cleaner-looking option is to install a wall-mounted Level 2 charger (Figure 3.1). These range in price from about $180 for aftermarket units to $500 for units from your EV manufacturer. According to HomeAdvisor, it will cost you $1,200 to $2,000 to have a Level 2 charger professionally installed [14]. This includes the parts and labor of installation, but doesn't include the charger itself.

FIGURE 3.1 A typical wall-mounted Level 2 charger.

Herr Loeffler/Shutterstock.com.

Level 3 provides even faster charging. It is also called DC fast charging. Level 3 requires a 480 V electric circuit, which is typically only found in commercially zoned or industrial zoned areas (Figure 3.2). It can usually recharge a battery up to 80% capacity in less than half an hour. The idea is that you plug in for about twenty minutes and grab a cup of coffee or snack or stretch your legs while you charge.

FIGURE 3.2 Look for signs like these. Many hotels now have charging stations for their guests and commercial charging stations are popping up along highways.

Courtesy of Chris Johnston and Ed Sobey.

azmanq/Shutterstock.com

Public charging stations are also becoming more common at grocery stores, libraries, and public and workplace parking lots (Figure 3.3). Routing apps show the locations of charging stations.

FIGURE 3.3 Free charging station.

Courtesy of Chris Johnston and Ed Sobey.

Cost to Charge Your EV

Compared to gas-powered vehicles, EVs are much cheaper to fill up. Charging at home takes advantage of the low cost of electric power. Most EV owners report a modest-to-unnoticeable increase in their electric bill. The average kilowatt rate was $0.1118/kW in the United States in 2021 [15]. This means that it will cost about $6.40 to charge an average-sized 50 kWh EV battery. Average monthly electric power bills in the US range from $100 to $165 per month [16].

If that EV with a 50 kWh battery gets a typical 200-mile range, it will cost about $2.80 to drive it 100 miles. As described above, a comparable gasoline-powered car will cost about six times more.

Many cities have free public charging stations. Some places to look are malls, shopping centers, and libraries. You can use the PlugShare app to find free EV charging stations.

FIGURE 3.4 Chart showing the fuel cost to add 100 miles of range to the average EV or internal-combustion vehicle. Provided by the US Bureau of Labor Statistics for the electricity rates and the US Energy Information Administration for the gas prices.

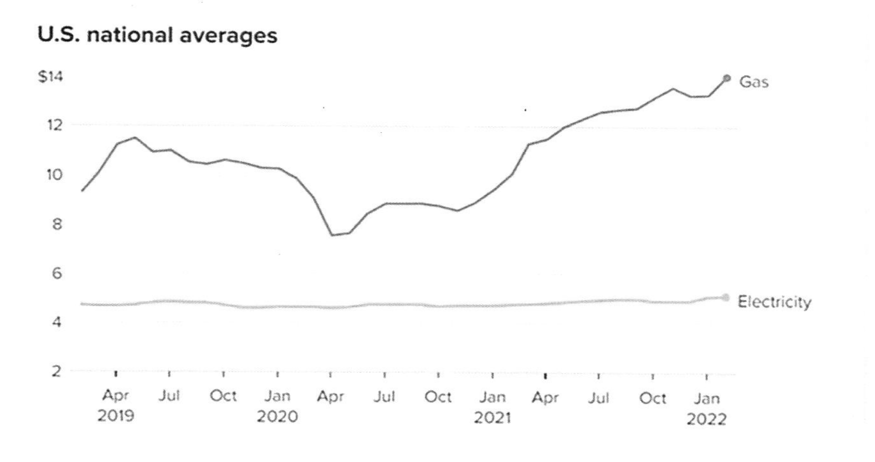

Electricity rates Courtesy of U.S. Bureau of Labor Statistics; Gas prices Courtesy of U.S. Energy Information Administration.

Tax Credits and Incentives

The federal government and some state governments offer tax credits, tax deductions, and other incentives that lower the cost of acquiring and operating an EV.

Internal Revenue Code (IRC) Section 30D provides a credit for Qualified Plug-In Electric Drive Motor Vehicles including passenger vehicles and light trucks. The largest tax deduction ($7,500) is offered by the US federal government. The credit begins to phase out when a manufacturer has sold at least 200,000 qualifying vehicles for use in the United States (determined on a cumulative basis for sales after December 31, 2009). Tesla was the first manufacturer to phase out followed by General Motors. For a complete index of vehicles by make, model, and manufacturing eligible for IRC 30D go to https://www.irs.gov/businesses/irc-30d-new-qualified-plug-in-electric-drive-motor-vehicle-credit [17].

Did you know that you can visit the US Department of Energy's Alternative Fuels Data Center's website to see laws and incentive programs listed by state; visit https://afdc.energy.gov/laws/state.

Eight states offer EV tax rebates, with Colorado offering the highest at $4,000. Many states and individual municipalities offer EV Supply Equipment grants, rebates, and other incentives. For example, Tucson Electric Power offers a rebate to residential customers that covers up to 75% of the cost of installing a home charger.

Warranty

US federal law mandates that EV battery warranties cover at least 8 years or 100,000 miles for the batteries of all EVs sold in the United States. However, the federal regulation only covers complete battery failure.

Excluding complete battery failure, companies offer a range of warranties. For example, BMW, Chevrolet, Nissan, Tesla (Model 3), and Volkswagen will replace a battery if it loses 30% to 40% of its capacity during the warranty period. Others may not. It is important to do your research before making a purchase.

Insurance

According to Consumer Reports, insurance premiums for EVs can be 16% to 26% higher than for comparable ICE cars. This is most likely because insurers don't have longstanding risk assessment data for EVs as they do for gasoline vehicles, which have been around for decades [18]. Over time, EV insurance rates should come down.

4

Common Misconceptions about EVs

Why do we have a chapter on common misconceptions and what's the origin of these misconceptions? There is a lot of misinformation about EVs. Some of it are old information that persist, and some are deliberately sowed to discourage people from purchasing EVs.

EVs Do Not Have Enough Range to Be Viable

Reality: Ten years ago, this was no myth. For example, in 2011, the Nissan LEAF was the first mass-market EV, and it had an effective range of 75 miles. The LEAF now has a range of 226 miles. The average range of the twenty-two mass-market EVs shipping in North America in 2022 is 284 miles [3]. The average range of a gasoline-powered car is about 275 miles.

EVs Are Not Greener than Gasoline- or Diesel-Powered Cars

Reality: Sometimes you can see a totally bogus claim with your eyes. Stopped at any intersection, you can see plumes of exhaust arising from ICE cars, especially those that need a tune-up. From the tailpipe of an EV, what do you see? You can't even see the tailpipe, let alone exhaust, because there isn't any.

Those exhaust plumes from ICE cars are composed of several greenhouse gases that we don't want to add to the atmosphere. Getting the gasoline from the ground into the tank of an ICE car uses more electricity than an EV uses in driving, that is, a gasoline-engine car that is sitting still has used more electricity than an EV will use to drive.

FIGURE 4.1 Clearly not an EV.

Tricky_Shark/Shutterstock.com.

Exploring for oil, pumping it out of the ground, shipping it, and refining it into gasoline and diesel is an energy-intensive process. EVs will continue to get greener as the power grid gets greener. ICE vehicles will remain dirty.

According to the US Energy Information Administration, US renewable electricity generation has doubled since 2008. Almost 90% of the increase in renewable energy came from wind and solar power generation. As of 2018, renewables provided 17.6% of electricity generation in the United States [19].

Meanwhile, ICEs are burning gasoline and diesel fuel and emitting into the atmosphere more than half of the total carbon monoxide and nitrogen oxides that humanity releases, and almost a quarter of the hydrocarbons (Figure 4.2) [20].

FIGURE 4.2 Annual CO_2 emissions of different vehicle types.

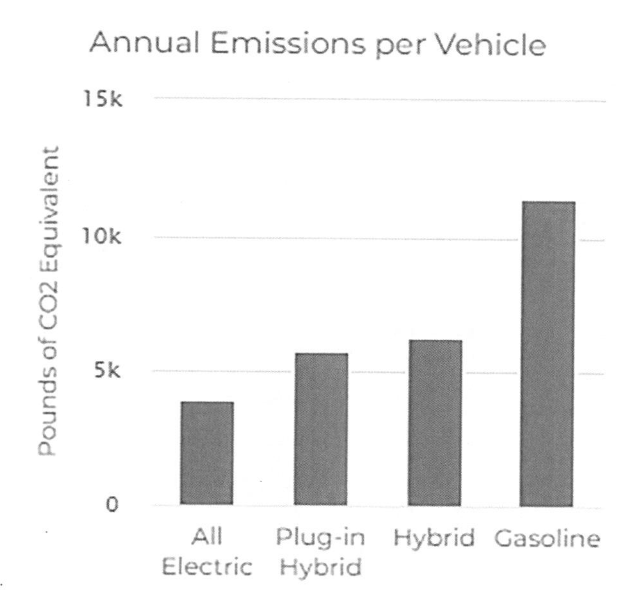

Courtesy of U.S. Department of Energy.

EVs Are Expensive to Maintain

Reality: EVs are much less expensive to maintain. The three big costs of operating a gasoline-powered vehicle are gas, oil changes, and brake replacement.

The cost of fuel for an EV is much lower—about a quarter as expensive as an ICE vehicle covering the same driving distance. The low cost of electric energy and the high efficiency of electric motors make EVs much cheaper to "fuel" than gasoline-powered vehicles. What does gasoline cost you? Compare that to about $0.75 per equivalent distance driven by an EV.

EVs don't require oil changes. Isn't that nice—not having to take your car in every few thousand miles to change the oil? Perhaps you crawl under your car in the driveway with a wrench in hand to do the job.

Because EV brakes last so much longer, they need to be replaced much less often. That saves a few hundred dollars. The brakes last longer because EVs use the motor to slow your forward motion, which takes some of the load off the brakes. This is called regenerative braking.

There Aren't Enough Public Charging Stations

Reality: The number of EV charging stations in the United States is growing rapidly. The Department of Energy reports that the United States now has over 20,000 EV charging stations, with more than 82,000 connectors. That's up from two years ago when there were about 16,000 public EV charging stations with about 43,000 connectors—an increase of 38% [21].

FIGURE 4.3 Internet map showing the location of charging stations in Seattle.

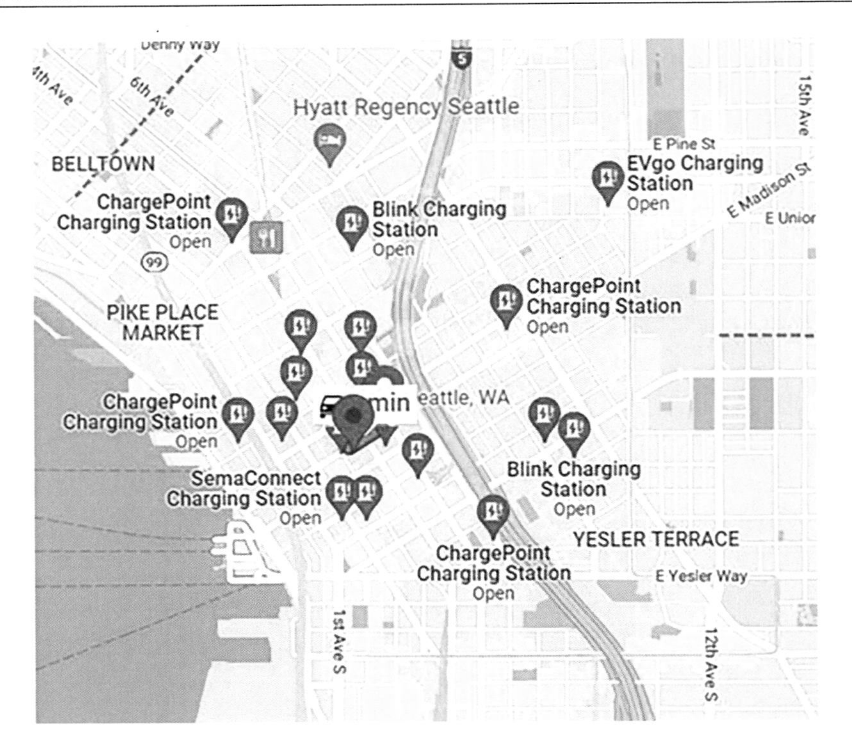

Courtesy of Chris Johnston and Ed Sobey.

Tesla is one of the larger providers of charging stations. They have 1,971 Supercharger Stations with 17,467 connectors, most positioned along major highways. The other big players are ChargePoint, with 30,000 connectors, and Electrify America, with 12,000 charging stations and 35,000 connectors [22].

In December 2021, a group of 50 power companies organized to install EV fast-charging stations along US highways by the end of 2023 [22].

There is a tendency to compare public EV charge stations with gas stations. Most EV owners charge their cars at home, which is not an option for ICE vehicles. So it is a bogus comparison to look at the number of gas stations versus the number of public charging stations when there are many thousands of residential charging stations.

EVs Are Worse for the Climate Because of Battery Manufacturing

An oft-mentioned narrative is that the carbon footprint to manufacture EV batteries is so high that it negates the environmental benefits of purchasing one. Let's dig into this by comparing the carbon footprint of a typical ICE vehicle versus a typical EV over a 200,000-mile life.

Reality: Over a 200,000-mile life, the average light-duty gasoline-powered vehicle has a carbon footprint that is more than two and a half times larger than the average light-duty EV (Table 4.1).

TABLE 4.1 Kilograms of CO_2 released from an ICE compared to an EV.

Kilograms of CO_2 released from...	ICE Light-duty vehicle	Battery electric Light-duty vehicle
Burning gasoline	69,182	0
Production of electricity	0	27,300
Production of gasoline	15,175	0
Manufacturing an EV battery	0	5,000
Totals	84,357	32,300

Courtesy of Chris Johnston and Ed Sobey.

According to the US Department of Energy, the average fuel economy for the model year 2020 light-duty vehicles increased to 25.7 miles per gallon [23]. Therefore, the average light-duty vehicle will consume 7,782 gallons in its 200,000-mile life. Following on, according to the US Environmental Protection Agency, a gallon of gasoline burned creates about 8.89 kg of CO_2, or **69,182 kg** after burning 7,782 gallons of gasoline [24].

Producing gasoline or diesel fuel is an energy-intensive process. A conservative estimate is that it takes at least 5 kWh of electricity to drill, transport, and refine a gallon of gasoline [25]. So 7,782 gallons of gas will require 38,910 kWh of electricity to produce or **15,175 kg** of CO_2.

A report from Brussels-based Transport and Environment think tank concludes that manufacturing an average EV battery has a carbon footprint of between **4,000 kg and 5,000 kg** of CO_2 emissions [26].

According to data from the US Department of Energy compiled by Eco Cost Savings, the average EV consumes 0.346 kWh per mile. So, over its 200,000-mile life, a typical light-duty EV will require 70,000 kWh [27]. Following on, according to the US Energy Information Administration, about 0.39 kg of CO_2 is emitted per kilowatt hour of electricity generated, or **27,300 kg** of CO_2 after using 70,000 kWh of electricity [28].

EV Batteries Don't Last and Will Cause a Recycling Problem

EVs on the market today use lithium-ion batteries. When Nissan started selling the first mass-produced EV in 2011, there were concerns about the LEAF batteries degrading over time. With more than ten years of experience, we know that these batteries lose about 1% of capacity every 18,750 miles, or less than 20% after 200,000 miles. Of course, the results will vary from manufacturer to manufacturer, but the general trend holds. Another point of confidence is that EVs are federally mandated to carry separate warranties for their battery packs for at least eight years or 100,000 miles.

Regarding recycling, gas-powered vehicles use lead-acid batteries. According to Battery Council International, lead-acid batteries have a recycling rate of 99.3% [29], making them the number one recycled consumer product in the United States. Lithium-ion batteries are made from more valuable metals and rare-earth elements, making them more likely to be commercially recycled. It is also worth noting that EV batteries typically won't go from the vehicle to the recycling plant because they will still have useful capacity. Many used EV batteries will have a post-vehicle life storing solar energy or wind energy, or in other power-grid applications.

The Electrical Grid Can't Support Millions of EVs

National Renewable Energy Laboratory research concludes that the existing grid can support the current EV charging load and the demand that would occur if 25% of the cars on the road were electric [30]. One reason is that a majority of EVs are charged when grid demand is low. It is also important to note that EVs won't reach 25% market share for some time. After that, grid-infrastructure upgrades can be made gradually and locally on a neighborhood-by-neighborhood basis. There won't suddenly be a need to upgrade the entire national grid infrastructure.

EVs Are Too Expensive

When it comes to comparing the price of EVs to comparable gas-powered cars, it depends on the market segment. For example, in the luxury midsize category, EVs are priced close to the price of comparable gas-powered cars. Some examples are as follows (Table 4.2).

TABLE 4.2 Cost comparison of gasoline-powered midsize luxury vehicle to an EV in the same class.

Luxury midsize category			
Gasoline powered		**Electric vehicles**	
Car	Starting MSRP	Car	Starting MSRP
BMW 330i	$42,300	Ford Mach-E	$43,895
Audi A4	$45,500	Tesla Model 3	$49,990

Courtesy of Chris Johnston and Ed Sobey.

In the economy market segment today, EVs are still priced at a premium, but you will need to factor in incentives. The US federal government and state governments offer tax credits, tax deductions, and other incentives that lower the cost of buying and operating an EV. The largest tax deduction is the $7,500 that is offered by the US federal government.

When it comes to filling the "tank" and maintenance, EVs are much less expensive. As we outlined in our "Cost of Ownership" section, an ICE vehicle will cost almost eight times more to operate and maintain than a comparable EV.

More Information

Information on cars and trucks is available at many locations on the internet. The federal government and consumer websites provide lots of background and information to compare vehicles.

We recommend our forthcoming book, *Electric Cars Have Arrived*, published by the SAE International. It expands on the information in this quick reference and has a car-buying guide for electric cars. Check it out online and at traditional bookstores.

About the Authors

Photo by John Gallagher.

Photo by John Gallagher www.bygallagher.com. (Ed on left, Chris on right).

Chris Johnston

Chris Johnston has decades of product management experience in telematics, mobile computing, and wireless communications including positions at Trimble Navigation, AT&T, Honeywell, and a couple of Silicon Valley startups. He also spent a year in India setting up an Internet-of-things practice for a major Indian corporation. Mr. Johnston has a B.S. in Electrical Engineering from

Purdue University and an MBA from Loyola University of Chicago. Chris lives in Washington State. When not working, he enjoys open water swimming, cycling, and flying (as a private pilot).

Ed Sobey

Ed Sobey holds a PhD in science and teaches for Semester at Sea. He also lectures at sea for passengers on several cruise lines and has traveled the equivalent of ten times around the world at sea. The Fulbright Commission has awarded Ed three grants for training science teachers in foreign countries. To date, he has trained teachers in more than 30 countries. He is a former naval officer and has directed five science centers, published 35 books, and hosted two television series on science and technology. Ed is a Fellow of The Explorers Club and has participated in two dozen scientific expeditions. He has conducted ocean research in winter in Antarctica, sailed across the Pacific Ocean in a small sailboat, and recorded whale sounds from an ocean kayak.

References

1. https://spectrumnews1.com/ca/la-west/transportation/2022/04/20/electric-vehicle-sales-surged-during-the-first-quarter
2. https://www.utilitydive.com/news/global-ev-sales-rise-80-in-2021-as-automakers-including-ford-gm-commit-t/609949/
3. https://ev-database.org/cheatsheet/range-electric-car
4. https://www.eia.gov/dnav/pet/pet_pri_gnd_dcus_nus_w.htm
5. https://www.eia.gov/electricity/monthly/epm_table_grapher.php?t=epmt_5_06_b
6. https://blog.evmatch.com/electric-vehicles-are-only-getting-cheaper-to-own/
7. https://www.energy.gov/sites/default/files/2020/12/f81/Bidirectional%20EVs%20%20Charging%20Roundtable%20Report.pdf
8. https://www.kbb.com/brake-repair/
9. https://www.kbb.com/oil-change/
10. https://crashstats.nhtsa.dot.gov/Api/Public/ViewPublication/809438#:~:text='%20Rollover%20crashes%20are%20more%20likely,been%20decreasing%20in%20recent%20years.
11. https://driveelectric.org.nz/consumer/what-is-an-ev/#:~:text=A%20fully%20electric%20car%20has,nearly%202%2C000%20in%20an%20ICEV!
12. https://www.geotab.com/blog/ev-battery-health/
13. https://www.fueleconomy.gov/feg/evtech.shtml
14. https://www.homeadvisor.com/
15. https://www.eia.gov/electricity/monthly/epm_table_grapher.php?t=epmt_5_06_b
16. https://www.eia.gov/electricity/sales_revenue_price/pdf/table5_a.pdf
17. https://www.irs.gov/businesses/plug-in-electric-vehicle-credit-irc-30-and-irc-30d
18. https://www.consumerreports.org/car-insurance/electric-vehicles-cost-more-to-insure-than-gasoline-powered-a6372607024/

19. https://www.eia.gov/todayinenergy/detail.php?id=38752#:~:text=U.S.%20renewable%20electricity%20generation%20has%20doubled%20since%202008&text=Renewables%20provided%2017.6%25%20of%20electricity,from%20wind%20and%20solar%20generation

20. https://www.epa.gov/greenvehicles/greenhouse-gas-emissions-typical-passenger-vehicle

21. https://www.energy.gov/eere/vehicles/articles/fotw-1174-february-22-2021-over-20000-new-electric-vehicle-charging-outlets

22. https://www.eenews.net/articles/major-u-s-utilities-plan-coast-to-coast-ev-charging-network/

23. https://www.energy.gov/eere/vehicles/articles/fotw-1177-march-15-2021-preliminary-data-show-average-fuel-economy-new-light#:~:text=Preliminary%20data%20for%20EPA's%202020,light%2Dduty%20vehicle%20fuel%20economy

24. https://www.epa.gov/greenvehicles/greenhouse-gas-emissions-typical-passenger-vehicle#:~:text=typical%20passenger%20vehicle%3F-,A%20typical%20passenger%20vehicle%20emits%20about%204.6%20metric%20tons%20of,8%2C887%20grams%20of%20CO2

25. https://greentransportation.info/energy-transportation/gasoline-costs-6kwh.html

26. https://www.brinknews.com/how-to-tell-whether-your-car-is-really-green/

27. https://ecocostsavings.com/average-electric-car-kwh-per-mile/#:~:text=The%20average%20electric%20car%20kWh%20per%20100%20miles%20(kWh%2F100,kWh%20to%20travel%201%20mile

28. https://www.eia.gov/tools/faqs/faq.php?id=74&t=11#:~:text=In%202020%2C%20total%20U.S.%20electricity,CO2%20emissions%20per%20kWh

29. https://batterycouncil.org/blogpost/1190989/289655/Study-Finds-Lead-Batteries-Are-Most-Recycled-Consumer-Product#:~:text=Study%20Finds%20Lead%20Batteries%20Are%20Most%20Recycled%20Consumer%20Product,-Posted%20By%20BCI&text=Today%2C%20in%20conjunction%20with%20America,recycling%20rate%20of%2099.3%20percent.

30. https://www.nrel.gov/news/features/2020/grid-coordination-opens-road-for-electric-vehicle-flexibility.html